Framework for implementing Integrated Vector Management (IVM) at district level in the South-East Asia Region

A step-by-step approach

Regional Office for South-East Asia

WHO Library Cataloguing-in-Publication data

World Health Organization, Regional Office for South-East Asia.
　　Framework for implementing integrated vector management (IVM) at district level in the South-East Asia Region: a-step-by-step approach.

1. Communicable Disease Control.
2. Disease Vector – prevention and control.
3. Insecticides.
4. Mosquito Control.
5. Pest Control.
6. South-East Asia.

ISBN　　978-92-9022-332-0　　　　　　　　(NLM classification: WA 110)

© World Health Organization 2008

All rights reserved. Requests for publications, or for permission to reproduce or translate WHO publications – whether for sale or for noncommercial distribution – can be obtained from Publishing and Sales, World Health Organization, Regional Office for South-East Asia, Indraprastha Estate, Mahatma Gandhi Marg, New Delhi 110 002, India (fax: +91 11 23370197; e-mail: publications@searo.who.int).

The designations employed and the presentation of the material in this publication do not imply the expression of any opinion whatsoever on the part of the World Health Organization concerning the legal status of any country, territory, city or area or of its authorities, or concerning the delimitation of its frontiers or boundaries. Dotted lines on maps represent approximate border lines for which there may not yet be full agreement.

The mention of specific companies or of certain manufacturers' products does not imply that they are endorsed or recommended by the World Health Organization in preference to others of a similar nature that are not mentioned. Errors and omissions excepted, the names of proprietary products are distinguished by initial capital letters.

All reasonable precautions have been taken by the World Health Organization to verify the information contained in this publication. However, the published material is being distributed without warranty of any kind, either expressed or implied. The responsibility for the interpretation and use of the material lies with the reader. In no event shall the World Health Organization be liable for damages arising from its use.

This publication does not necessarily represent the decisions or policies of the World Health Organization.

Printed in India

Contents

Page

Glossary of terms v

1. Introduction 1
2. Policies in Support of Vector Control 3
3. District IVM Planning Cycle 5
   3.1 Purpose of developing a framework on district IVM plan 5
   3.2 Definition of a district 6
   3.3 Review of policy framework and institutional capacity 6
   3.4 Local situation analysis and needs assessment 8
   3.5 Setting goal and objectives 11
   3.6 Implementation process 12
   3.8 Budgeting 28
4. References 28
5. List of Contributors 29

Glossary of terms

CVC: Comprehensive Vector Control

FAO: Food and Agriculture Organization of the United Nations

GIS: Geographical Information System

FFS: Farmer Field School

HIA: Health Impact Assessment

IEC: Information, Education and Communication

IPM: Integrated Pest Management

ITN: Insecticide Treated Nets

ITM: Insecticide Treated Material

IVC: Integrated Vector Control

IVM: Integrated Vector Management

IPVM: Integrated Pest and Vector Management

KABP: Knowledge, Attitudes, Beliefs and Practices

LLIN: Long Lasting Insecticidal Net

SVC: Selective Vector Control

VBD: Vector-Borne Disease

VBDCP: Vector-Borne Disease Control Programme

VMNA: vector management needs assessment

WHO: World Health Organization

1. Introduction

Major vector-borne diseases (VBD) occurring in the South-East Asian Region are malaria, dengue, Japanese encephalitis, filariasis and visceral leishmaniasis (Kala-azar). Recently chikungunya has re-emerged in large parts of the Region. Not only do these VBD affect health adversely, they also make people poorer and impede overall human development. Poverty alleviation and the use of multisectoral approaches for sustainable development to reduce the burden of VBD assume great significance in reducing the overall burden of communicable diseases.

Human-made ecological changes, which may be caused by developmental activities and expansion of agriculture, can alter environmental dynamics, increase community vulnerability, and enhance the risk of transmission of VBD. Conventionally, the control of disease vectors has mainly relied on the use of chemical insecticides. The failure to effectively reduce the burden of VBD has arisen from a number of factors – human, technical (including insecticidal and drug resistance), operational, ecological, and economic. Current approaches to controlling different VBD work in near isolation from each other. In certain situations, opportunities exist for optimum control of vectors of two or more diseases to be organized together and managed effectively by optimal use of available technologies, interventions and improved resources at a local level, taking into account health sector reforms wherever possible. This requires an ecosystem approach to vector control at a decentralized level, in particular at the district and community levels.

Potential of vector control

Most South-East Asian countries have fairly long experience of using vector control for the prevention and control of malaria. Vector control can play a key role in prevention and reduction of disease transmission in one or more of the following ways: by reducing the vector density or abundance, reducing longevity, and preventing human-vector contact. The potential of vector control can be exploited in reducing the spread of drug resistance, preventing/managing epidemics, and lowering the risks of re-emergence or the introduction of diseases in new or low endemic areas. Although effective use of vector control led to the reduction/eradication of VBD in large parts of the world during the twentieth century, its potential is not being fully exploited at present. Vector control is the sole method of prevention against dengue and chikungunya, and a major means of controlling Japanese encephalitis, malaria and Kala-azar. It can play a complementary role in the control of filariasis. The collateral benefits of methods such as environmental management, household sanitation and insecticide treated nets (ITN)/long-lasting insecticidal nets (LLIN) extend to controlling insects of public health importance such as houseflies, cockroaches, rodents, bed bugs etc. There is, therefore, a need to increase access to effective preventive and vector control interventions in areas that are at high risk of vector borne diseases.

Progress toward an Integrated Vector Management (IVM) approach

The re-emergence of diseases and problems triggered by the excessive dependence on insecticides led the World Health Organization (WHO) to recognise the importance of **integrated vector control (IVC)** in the early 1980s along the line of integrated pest control used in agriculture. The IVC approach was described as the "utilization of all appropriate, safe and compatible means of control to bring about an effective degree of vector suppression in a cost-effective manner" (WHO 1983). It was essentially a unified plan that included use of one or more compatible methods of control.

Later, **selective vector control (SVC)** evolved as the "application of targeted, site-specific and cost-effective activities to reduce malaria morbidity and mortality" (WHO 1992, 1995). It required delineation of areas requiring vector control and then selection of methods based on knowledge of vectors and environmental, ecological, social, economic and health-service features.

Since it was considered that a unified management structure could manage two or more VBD prevalent in the same area, the concept of **comprehensive vector control (CVC)** was referred to as "control of the vectors of two or more co-prevalent diseases through a unified managerial structure using similar or different methods".

Integrated Vector Management (IVM), the most recent approach to vector control, uses sound principles of management and allows full consideration of the determinants of disease transmission and control. The latest definition of IVM proposed by WHO in 2007 describes it as "a rational decision-making process for the optimal use of resources for vector control."

IVM is a decision-making process for the management of vector populations, so as to reduce or interrupt transmission of vector-borne diseases. Its characteristic features include:

- selection of methods based on knowledge of local vector biology, disease transmission and morbidity
- utilization of a range of interventions, often in combination and synergistically
- collaboration within the health sector, researchers and with other public and private sectors that impact on vector breeding
- engagement with local communities and other stakeholders
- a public health regulatory and legislative framework
- rational use of insecticides
- good management practices

An IVM approach takes into account the available health infrastructure and resources and integrates all available and effective measures, whether chemical, biological, or environmental. IVM also encourages an integrated approach to disease control.

2. Policies in Support of Vector Control

A number of international policies have lent support to the importance of vector control in prevention and control of diseases. The Global Malaria Control Strategy (1992) included planning and implementing selective and sustainable preventive measures, including vector control, as one of its four basic technical elements.

The Fiftieth World Health Assembly (in resolution WHA 50.13, 1997[1]) urged steps to reduce reliance on insecticides for control of VBD through the promotion of integrated pest-management approaches in accordance with WHO guidelines, and through support for the development and adaptation of viable alternative methods of disease vector control. It further suggested to reduce reliance on insecticides through the promotion of an integrated approach. This Resolution is in line with the commitments of the Parties of the Stockholm Convention on Persistent Organic Pollutants[2], and the Bahia Declaration of the Intergovernmental Forum on Chemical Safety[3].

The commitments of SEAR countries made under UN Agenda 21[4] and World Summit on Sustainable Development to minimize significant adverse effects of chemicals on human health and the environment, require the implementation of environmentally sound vector control strategies.

Since the development of the Global Malaria Control Strategy, three regional consultations have specially discussed vector control needs in the South-East Asia (SEA) Region: an intercountry workshop in Bangalore, India that recommended development of policies and action plans for planning and implementation of vector control for malaria (1995); a regional consultation on disease vector surveillance and control at ports and airports held in Bangkok, Thailand (1998); and an intercountry consultative meeting held in Myanmar to discuss alternative approaches to vector control (1999).

In 2004, WHO issued the new Integrated Vector Management (IVM) strategy[5]. IVM is based on the premise that effective control is not the sole preserve of the health sector but requires the collaboration of various public and private agencies and community participation. The active engagement of communities is a key factor in assuring sustainability. IVM entails the use of a range of biological, chemical and physical interventions of proven efficacy, separately or in combination, in order to implement more cost-effective control and reduce reliance on any single intervention. In 2006, the WHO Regional Office for South-East Asia organized a regional workshop in Puducherry, India to promote the implementation of IVM in the SEA Region.[6]

---

[1] www.who.int/ipcs/publications/wha/whares_53_13/en/index.html
[2] www.pops.int/
[3] www.who.int/ifcs/documents/forums/forum3/en/Bahia.pdf
[4] www.un.org/esa/sustdev/documents/agenda21/index.htm
[5] http://whqlibdoc.who.int/hq/2004/WHO_CDS_CPE_PVC_2004_10.pdf
[6] See report at : http://www.searo.who.int/EN/Section23/Section1001/Section1110_13215.htm

To advocate and help implement the IVM approach in Member countries, a SEA Regional Framework (WHO Regional Office for South-East Asia, 2005) for IVM was developed. WHO Regional Office for South-East Asia is supporting the implementation of the Revised Malaria Control Strategy (2006-2010), which was endorsed at the 2007 Regional Committee Meeting.[7] The Revised Malaria Control Strategy of SEA Region includes the application of IVM and promoting the development of healthy public policies (WHO Regional Office for South-East Asia, 2006)[8]. More recently, the World Health Assembly (in Resolution 60.18, 2007) has called for support in developing country capacities in expanding the use of effective interventions for malaria control including the use of an IVM approach. It is now time for Member countries to adopt and effectively implement IVM.

3. District IVM Planning Cycle

3.1 Purpose of developing a framework on district IVM plan

It is recognized that the technical, managerial and operational capacities of district-level managers are inadequate or weak and can jeopardize the planning and implementation of IVM at district and sub-district levels. Non-health sectors are usually not aware of the importance of their roles in vector management, and the operational and managerial mechanisms for taking synergistic actions are thus fragile. Guidance at district level is therefore urgently required to address the appropriateness of intervention methods for an integrated programme. This framework, which should not be seen as prescriptive, provides generic guidance to develop and implement an IVM approach at district or equivalent levels, as well as to monitor and evaluate its impacts. It is designed mainly for use by the VBD control programme managers and other concerned stakeholders at the district level and below, but also for reference of state and national level policy- and decision-makers.

---

[7] (see: http://www.searo.who.int/LinkFiles/Provisional_Agenda_SEA-RC60-9_Agenda_Item_11.pdf)
[8] *Healthy public policies* improve the conditions under which people live: secure, safe, adequate, and sustainable livelihoods, lifestyles, and environments, including housing, education, nutrition, information exchange, child care, transportation, and necessary community and personal social and health services. Policy adequacy may be measured by its impact on population health.

3.2 Definition of a district

A district is regarded as an administrative sub-division in a province or state in a country. However, this would vary from country to country depending on the size and population. In this circumstance, country specific sub-divisions should be adhered to.

3.3 Review of policy framework and institutional capacity

A first step to develop a district IVM plan is to review the policy framework that guides suitable actions for implementing IVM. Many of the policy issues listed below may not be under the purview of the district-level manager, yet their assessment will be useful for consideration by higher-level policy- and decision-makers. This exercise should be conducted by the focal point for vector control and her/his staff. The various key stakeholders, including community representatives, identified during this process should be brought on board, integrated into the "IVM team" and given responsibilities in completing the analysis.

Review of the Institutional framework

- ➢ Provision of an enabling environment
    - Review the current health system's capacity at district level for IVM implementation.
    - Establish strengths and identify gaps in terms of human, technical and financial resources.
    - Develop a plan to address the gaps identified, including the appropriate recruitment and career policies of vector-control cadres and procurement policies for supplies.

Health/VBD control programme policies

- ➢ Is IVM policy a component of provincial/state and district health policies?
- ➢ Are there policy guidelines on the use of IVM interventions? For example:
    - use of larvivorous fish, biolarvicides, insect growth regulators;

- distribution, management and use of ITN or LLIN;
- application of chemical larvicides, insecticides for indoor residual spraying and space spraying, and environmental management methods;
- production, import/export, storage, distribution, and use of ITN/LLIN;
- safe disposal of worn-out ITN/LLIN, empty/used insecticide containers and unwanted/obsolete insecticide stocks;
- identification of community stakeholders for active community participation in IVM.

Environmental policy

- Are there specific provincial/state laws, legislations or operational guidelines on the management of pesticides for public health? For example:
    - to regulate the import of insecticides;
    - to grant license to manufacture, sell, exhibit for sale or distribute any insecticide, including safety measures necessary to prevent risk to human beings or animals in manufacture, sale, storage, transport and distribution;
    - use of pesticides and disposal of unwanted/obsolete public health pesticides and empty containers and bags;
- Does a policy exist to incorporate health impact assessment in development particularly related to vector borne diseases? For example:
    - water resource development projects
    - agriculture development projects
    - industrial and infrastructure development projects
    - rainwater harvesting programmes
    - urban development planning
    - rural development programmes

Agriculture policies

- What are the policies related to the management of pesticides, notably concerning storage and use of pesticides in agriculture, and disposal of used containers/bags?
- Are there policies and programmes to reduce occupational exposures of farmers using pesticides?
- Are there integrated pest management (IPM) programmes and/or farmer education programmes such as Farmer Field Schools[9] being implemented? If so, can vector management be integrated into them?
- Is there any policy proposed to incorporate water management/intermittent irrigation system in rice fields to control the breeding of mosquito vectors, especially with reference to Japanese encephalitis and malaria?

Financial and trade policies

Study and analyze the current policies and issues related to taxes and tariffs/financial incentives/subsidies for the importation, manufacture and sale of insecticides, pesticide equipment, ITN and/or LLIN and other inputs used in vector control. Identify strengths and weaknesses and develop proposals for improving efficiency and sustainability.

3.4 Local situation analysis and needs assessment

Epidemiological and entomological assessment

A great number of local factors determines disease transmission in a given eco-epidemiological setting, such as parasite prevalence, vector species and their bionomics, human behaviour, climate (temperature, humidity, rainfall), terrain/physiography, etc. Local situation analysis is the key to identifying the major determinants of disease transmission and to establishing the needed response capacity and capability of the health

---

[9] www.farmerfieldschool.net/document_en/FFS_GUIDe.doc

system. The local epidemiological and entomological assessment, done by the "IVM team", includes:

- identification and mapping of main eco-epidemiological settings in the district, e.g. irrigated areas, project development areas, urban settings, etc.;
- identification of priority VBD and assessment of the disease burden (mortality, morbidity, economic impact, status of drug resistance) using data collected by the routine surveillance system or through special surveys/studies;
- identification and mapping of major determinants of VBD and communities at risk;
- entomological assessment, which includes:
  - Review available data on biology, ecology and behaviour of relevant vector species.
  - Conduct rapid entomological surveys to determine distribution, and prevalence of vector species if sufficient data are not available.
  - Determine sites of disease transmission to confirm indigenous transmission versus importation of cases.
  - Determine the status of insecticide susceptibility of local vectors.

Review of pesticide management practices

VBD programmes often use insecticides, which when not transported, stored and used according to internationally accepted safety guidelines, can cause occupational and environmental risks. As part of the situation analysis, collect the following information:

- How are insecticides transported?
- Where are they stored, i.e. in secured places, under ambient temperature/sunlight conditions, away from human dwellings?
- What is the mechanism of disposal of used insecticide containers and bags?

> Are sprayers provided with training and with protective clothing?

> What is the mechanism of medical follow-up, if warranted, and attendance in case of medical emergency?

> How are unwanted/obsolete pesticides stocks managed?

Assessment of challenges and constraints

Planning cycles usually make an assessment of the possible risks that might impede successful implementation of the intended interventions. Some of these may be:

> Inadequate technical and programme management capacities in IVM;

> Diminishing discipline of trained medical entomologists;

> Diminishing resources leading to grossly inadequate access and coverage, and burden of vectors created by development sectors passed on to health sector to mitigate;

> Spread of insecticide resistance;

> Poor pesticide management practices;

> Limited arsenal of safe, effective and cost-effective insecticides;

> Increase in areas with enhanced environmental receptivity requiring greater demands for vector control on already overstretched vector control services;

> Application of vector control in decentralized health system;

> Vector control confined to only public health system while private pest control operators work in isolation;

> Lack of synergy between health and agriculture, which use pesticides that have implications for the effectiveness of public health insecticides;

> Low visibility of vector control/IVM in health sector reform.

Vector Management Needs Assessment

Implementation of IVM requires a thorough situation analysis and assessment of challenges and specific needs as explained above. This is followed up by a vector management needs assessment (VMNA). Possible approaches to conducting VMNA are suggested below.

- assessing needs to build a health public policy framework for preventive actions in development for better health and a safe environment;
- assessing institutional building needs to strengthen health system infrastructure and capacity;
- technical needs assessment for the selection of interventions;
- assessing decision-making or programme management capacity to plan, implement, monitor and evaluate IVM;
- assessing needs for multisectoral collaboration;
- need to solicit community participation in vector control;
- need for standard tools and guidelines on IVM implementation;
- human and financial resource needs;
- human resource development needs.

3.5 Setting goal and objectives

Goal:

The overall goal of the district-level plan should be the same as for the overall reduction of the burden of VBD. Thus, it should "contribute to the reduction of targeted vector borne disease(s) at district level" compared with a baseline level.

Specific objective:

To reduce the elements of disease transmission: reduce breeding and abundance, vector longevity and human-vector contact.

General objectives:

- to strengthen technical and managerial capacity for the epidemiological situation analysis, vector control needs assessment, and planning, implementing and monitoring of the IVM implementation processes;
- to assist with the development and implementation of IVM strategy at the district and sub-district levels together with other partners and sectors, making use of synergistic effects;
- to facilitate situational analysis and decision-making by local partners for IVM strategies at the municipality and community levels;
- to document and publicize the process and outcomes;
- to develop synergies/coordinated action for sustained vector management within and outside of the health sector;
- to increase visibility of IVM for its replicability and sustainability.

3.6 Implementation process

Problem-solving in IVM

Implementation of IVM requires assessment of possible challenges and obstacles. Addressing challenges and strengthening health services according to IVM needs will require a favourable policy environment and remedial actions such as the following:

- adoption of IVM as part of the strategies at national, provincial/state and district levels;
- seeking political commitment at district level to create an enabling policy environment in favour of IVM;
- strengthening institutional framework to facilitate multisectoral collaboration and joint planning;
- implementing regulatory legislation;
- policy to incorporate health impact assessment in development;
- encouraging other sectors to incorporate IVM in sectoral budgets and participate in joint activities in an integrated collaborative approach;

- decentralized health system providing greater opportunities to localize analysis and decision-making on vector control with participation of local communities;
- mobilizing resources for increasing access and rational use;
- strengthening district-level core capacity and capabilities;
- setting-up of a core IVM team at district level with multidisciplinary expertise, namely entomological and vector control, epidemiological, environmental, sociological, programme management skills and the skills required to implement activities by local actors and communities,
- orienting the IVM team on strategic functions to implement IVM in the district.

The IVM team will be required to perform a number of functions, such as:

- organize epidemiological situation analysis and stratification of areas/communities according to risk of disease;
- develop necessary local-level capacity within and outside the health sector;
- make decisions together with other relevant actors at the district level on selecting strategic interventions and implementing them;
- coordinate multisectoral actions;
- monitor implementation processes and evaluate impacts;
- guide the participation of stakeholder communities by selecting prime movers, and community health workers, and empower them to undertake local situation analysis and evidence-based management decisions through a participatory approach to deal with local situations.

Selection of priority diseases for integrated action

Local decisions are required to be taken on the priority vector borne diseases to be managed using an IVM approach. This will however depend on a number of factors such as commonalities in vector ecology, biology and behaviour, amenability to interventions, and so on.

Examples include:

- control of malaria (vector: *Anopheles stephensi*), Japanese encephalitis (vector: *Culex tritaeniorhynchus*) and dengue (vector: *Aedes aegypti*) in certain urban settings;

- control of urban malaria (vector: *Anopheles stephensi*), lymphatic filariasis (vector: *Culex* species), dengue and chikungunya (vector: *Aedes* species) in certain cities of India;

- control of rural malaria and visceral leishmaniasis (Kala-azar) coexisting in some areas of Bangladesh and India;

- control of dengue and chikungunya in rural areas with malaria control;

- control of vectors of malaria and Japanese encephalitis breeding in paddy fields, etc.

Stratification of risk area

The stratification of the risk areas and communities is a generic process that guides decision-making for rational use of appropriate interventions in place and time and judicious use of resources to achieve maximum impact on the targeted disease(s). The stratification should be first done at a macro level to classify and confirm major ecotypes, establish relative importance of different ecotypes to malaria burden and set priorities. Disease burden is often stratified as holo-, hyper-, meso- and hypo-endemic, or with high, medium and low incidence depending upon mortality and morbidity rates and level of drug resistance. Low-incidence areas may be prone to sudden increase in risks and assume importance since preventive interventions may be required against re-introduction/re-emergence of the disease.

Micro-stratification at municipality, community, village and household levels requires detailed information to be acquired through a number of sources, i.e. through the existing surveillance systems, research institutions, documented records and special investigations. Members of the local community will be a key source of data, both in terms of disease burden and in terms of indigenous knowledge. The computer-aided geographical information system (GIS) and use of portable geographical positioning systems are useful in stratification. The GIS is very commonly used today for disease risk mapping but requires investment both in infrastructure and training of personnel.

In a decentralized IVM programme, however, micro-stratification is conducted through participatory mapping exercises involving local partners and communities, which is carried out as part of their situation analysis. This forms the basis for planning their local IVM programmes. On completion of micro-stratification, determine where there is need for vector management and for which epidemiological stratum, particularly in:

➢ reduction of transmission in high-risk areas;

➢ control of disease endemicity;

➢ prevention and control of epidemics;

➢ drug/insecticide resistance management;

➢ elimination of new foci of infection in disease free areas;

➢ prevention of re-introduction of disease in low endemicity areas.

If there is a potential role for vector management, identify vector(s) in each stratum. For each vector implicated, determine breeding sites, vector behaviour (feeding and adult resting sites), history of insecticide resistance and vector ecology. Based on these parameters, determine which method(s) of vector control is suitable.

Box 1: *Farmer Field School Approach in IPVM*

Agriculture and livestock-raising are important subsistence activities in South-East Asia. The link between malaria and irrigation/agriculture development has been known for a long time. Rice agro-ecosystems also support vectors of Japanese encephalitis in some countries. Among the main type of farming systems, modern agriculture is dependant on irrigation and inputs of fertilizers and pesticides. Against the wisdom of using traditional farming practices, excessive use of pesticides in agriculture has resulted *inter alia* in (a) disturbance in the natural balance between harmful crop pests and beneficial predators in local ecosystems, (b) increased occupational exposures of farming community and incidence of pesticide poisoning, (c) pesticide residues in food and environment, and (d) development of insecticide resistance of mosquito vectors.

Integrated Pest Management

Realizing the ill effects of excessive dependence on chemical pesticides, experimentation on the Integrated Pest Management (IPM) approach began in the late 1970s in the Philippines. IPM promotes the philosophy of "healthy fields for healthy people" and is a farmer-based agro-ecosystem management approach that uses broadly four principles: grow healthy crops using good agricultural practices, conserve natural enemies to reduce the use of chemical pesticides, conduct regular field observations to make timely and evidence-based management decisions, and empower farmers to become experts through a participatory learning approach to deal with local situations.

Farmer Field School Approach

Over the years, IPM led to the development of the Farmer Field School (FFS) movement, with major advancement in Indonesia in the late 1980s[1]. Unlike traditional farmer education, FFS is a modern, participatory learning and community empowering approach based on season-long practical demonstration of improved farming practices to protect the farming community and the agro-ecosystems from the ill effects of pesticide use, thereby aiming to create sustainable agriculture and environment. The FFSs, comprising groups of 15-30 farmers, are facilitated by agricultural extension trainers, from land preparation right up to the harvesting stage. Farmers participate in weekly learning cycles throughout the crop-season when they collectively learn to conduct agro-ecosystem analysis, identify agricultural pests and beneficial predators, and make informed decisions about crop management and use of pesticides. They compare processes and outcomes in the IPM plots with those in the non-IPM plots.

The IPM FFS approach was applied in rice cultivation in the mid-1990s in Asia and later expanded to vegetables, cotton and various other crops in other regions. A wide range of benefits, including development impacts, have been reported of IPM FFS in a recent evaluation of 25 IPM studies in Asia (Bangladesh, China, Cambodia, Indonesia, Philippines, Sri Lanka, Thailand, Vietnam) and other regions[2]. India has a large IPM programme, including the rice IPM.

The IPM FFS programme in Sri Lanka started in 1995. External evaluations[3] in 2002 reported that farmer practices resulted in substantial increase in yield and profits from rice cultivation, and reduced insecticide use causing savings in agrochemical inputs. IPM was found to be cost-effective, providing motivation, cooperation and a sense of programme ownership to farming communities. These positive FFS experiences have led to the integration of vector management with IPM leading to an Integrated Pest and Vector Management (IPVM) project since 2002.The IPVM strategy has helped farmers to reduce health risks associated with vector-borne diseases and pesticides. In its current phase, the project is expanding the curriculum to cover the health effects of pesticides and enhancing preventive community action and personal protection by participating in surveillance activities. The Malaria Control Programme plans to adopt the integrated pest and vector management strategy to prevent malaria in areas of low transmission since there is an additive effect between the use of mosquito nets and the strategy4. In addition to its suitability under Sri Lankan conditions, the integrated pest and vector management approach is potentially replicable in other countries and other regions – it could be used as an interdisciplinary topic for education in secondary schools.

---

[1] Pontius J, Dilts R, Bartlett A (eds). *Ten Years of Building Community: From Farmer Field Schools to Community IPM*. FAO Community IPM Programme: Jakarta, 2000.

[2] van den Berg, H (2004). IPM Farmer Field Schools: A synthesis of 25 impact evaluations. Report prepared for the Global IPM Facility, Wageningen University, January 2004.

[3] Henk van den Berg H, Senerath H, Amarasinghe H. *Participatory IPM in Sri Lanka: a Broad-Scale and an In-Depth Impact Analysis*. A Report prepared for the FAO Programme for Community IPM in Asia, Jakarta, 2002.

[4] Evaluation Report of the Integrated Pest and Vector Management (IPVM) Project in Sri Lanka- Mission Report", WHO/SEARO, New Delhi, 2006

Making decisions on intervention methods

There might be several possible approaches to managing vectors. In situations in rural areas where the agriculture sector has no direct role to play, vector control will need to be organised through an IVM approach. Where opportunity exists to work with the agriculture sector (see Box 1), apply an integrated pest and vector management (IPVM) approach, for example in control of rice breeding mosquito vectors of malaria (e.g. *An. culicifacies* in Sri Lanka and India, *An. sinensis* and *An. anthropophagus* in the Democratic People's Republic of Korea, *An. aconitus* in Indonesia) and/or Japanese encephalitis (*Cx. tritaeniorhynchus*) in rural areas of South-East Asian countries. In urban areas, communities can participate in sanitation efforts in the peri-domestic environment. Other forms of IVM are those involving the industrial and project sectors such as brick making, the construction sector (roads, buildings), etc. The major vectors of malaria and other vector borne diseases are listed in Table 1.

Table 1: Distribution of principal vectors of diseases in the WHO South-East Asia Region

Disease	Country	Vectors
Malaria	Bangladesh:	*An. dirus, An. minimus, An. sundaicus* and *An. philippinensis*
	Bhutan:	*An. minimus, An fluviatilis* and *An dirus*
	DPR Korea	*An. sinensis* and *An. anthropophagus*
	India:	*An. culicifacies, An. dirus, An. fluviatilis, An. minimus, An. stephensi* and *An. sundaicus;* secondary vectors: *An. annularis* and *An. philippinensis.*
	Indonesia:	*An. aconitus, An. balabacensis, An. barbirostris, An. farauti, An. maculatus, An. minimus* and *An. sundaicus;* secondary vectors: *An. fluviatilis, An. karwari, An. koliensis, An. letifer, An. leucosphyrus, An. nigerrimus, An. punctulatus, An. subpictus, An. umbrosus;*
	Maldives:	No indigenous transmission of malaria occurs in Maldives and the earlier known vector species, viz. *An. tesselatus* and *An. subpictus,* are no longer found there.

Disease	Country	Vectors
	Myanmar:	An. dirus, An. minimus, An maculatus, An. sundaicus, An. philippinensis and An. annularis (Rakhaine state)
	Nepal:	An. annularis, An. fluviatilis and An. maculatus
	Sri Lanka:	An. culicifacies and An. subpictus
	Thailand:	An. dirus, An. minimus, An maculatus, An. sundaicus and An. aconitus
	Timor-Leste:	An. barbirostris
Dengue/ Chikungunya		Aedes aegypti and Aedes albopictus (secondary vector)
Japanese encephalitis		Culex tritaeniorhynchus, Culex vishnui, Culex pseudovishnui, Culex gellidus and Culex fuscocephalus
Lymphatic filariasis		Bancroftian filariasis and Culex quinquefasciatus Bugian filariasis: Mansonia spp
Visceral Leishmaniasis		Phlebotomus argentipes

Selection of IVM components is based on situation analysis, technical effectiveness and cost-effectiveness of the methods, acceptance by communities, availability of resources, environmental safety and feasibility. The intervention will generally have one or more of the following actions to reduce vectorial capacity:

Objective	Action	Method
Reduce vector abundance	Reduce the number of sites where vector larvae grow	Environmental management
	Reduce number of larvae or prevent insects from reaching adult stage	Larvivorous fish, bio-larvicides, insect growth regulators, and other parasites, chemical larvicides
	Kill adult insects when they rest on sprayed surfaces	Indoor residual spraying (IRS)
	Kill insects as they alight on treated surfaces, repel them	Insecticide-treated materials (ITM) such as mosquito

Objective	Action	Method
	or inhibit them from feeding/biting	nets, curtains and *chaddars* (cloth sheets)
	Reduce the population of adult insects	Space spraying in urban areas
Reduction of vector survival and longevity	Reduce life of the insect before it reaches infective age	IRS and ITM
Reduction of human-vector contact	Reduce opportunities for insects to enter the house	House improvements such as screens with ITM
	Repel insects before they bite	Repellents: coils, mats, lotions, cream, vaporizers

The currently available tools and approaches for controlling disease vectors and pests of public health importance are further mentioned in Table 2. A range of these interventions can be applied either alone or synergistically in a variety of situations. For example,

- IPVM - in this case there may be one or more vector borne diseases and at least two sectors will be working in synergy;

- control of a single vector borne disease in a given area using more than one vector control method;

- control of two or more vector borne diseases in a geographical area or community where a single intervention might be preferred, such as environmental management and social mobilization for control of dengue and chikungunya; or control of dengue and urban malaria vectors; or control of dengue, chikungunya and malaria vectors in an urban setting; or control of Kala-azar and malaria using indoor residual spraying;

- a common vector control method against vectors of two different diseases in two separate areas/communities.

Table 2: Methods for control of disease vectors and pests of public health importance. The delivery and individual/combined effectiveness of these methods depend on local situations and factors of vector biology and behaviour.

Disease/ pests	Main vector(s)	Chemical control					Biological control		Environ-mental management[4]	Legislation
		IRS	ITNs/LLINs/ITM	Repellents	Larviciding	Space spraying	Larvivorous fish	Biolarvicides		By laws, regulations
	Effects on elements of vectorial capacity →	1, 2, 3	1, 2, 3	3	1	1, 3	1	1	1	Supportive regulation
Malaria	Anopheles spp.	+	+	+	+	+	+	+	+.	+
Dengue	Aedes aegypti, Ae. albopictus		+	+	+	+		+	+	+
Chikungunya	Aedes aegypti		+	+	+	+		+	+	+
Japanese encephalitis	Culex vishnui gp.	+	+	+		+			+	+
Leishmaniasis, Chandipura virus	Phlebotomus spp.	+	+						+	
Filariasis	Cx. quinque-fasciatus, Ma. uniformis, annulifera, indiana	+	+	+	+		+	+	+	+
Household pests:										
Houseflies	Musca domestica	+				+			+	+
Cockroaches	Periplanata americana	+, 5, 6							+	
Bedbugs	Cimex lectularius	+, 5	+						+	
Fleas	Xenopsylla cheopis	5							+	+
Rodents	Various spp.	7							+	

[1] Reduce abundance

[2] Reduce survival/longevity

[3]Reduce human-vector/pest contact

[4]Modifications: filling, drainage, impounding, etc.

Manipulation: changing water level, intermittent drying, flushing, water salinity changes, stream-bank shading, weeding, vegetation clearance, deliberate water pollution, mosquito-proofing containers, enforcing dry-pot day; Other methods: zooprophylaxis, solid waste management, house proofing/siting, piggeries location, use of polystyrene beads etc.

[5]Insecticide dusting of bottom of walls/infested sites

[6]Insecticide baits

[7]Rodenticides

Criteria and procedures for selection of intervention tools

The broad categories of proven vector control methods are listed below.

(1) Chemical control, such as by IRS, use of ITN/ITM/LLIN, larviciding, space spraying (fogging) and repellents. In the SEA Region, application of IRS is the mainstay in two principal situations: in intense malaria transmission areas against endophilic vectors, and in prevention/control of malaria epidemics. WHO promotes effective and safe use of insecticides for IRS. IRS requires adequate skills within the health system and is an intensive activity requiring detailed planning and supervision. The use of ITN/LLIN is recommended in malaria endemic areas especially to protect children, expectant mothers, and socio-economically vulnerable and unreachable communities. Their acceptance will be high in communities traditionally using the nets. They are also preferred in situations where it is difficult to implement IRS operationally.

(2) Environmental management is preferred in situations where mosquito breeding places are well defined and amenable to such an approach, for example in areas with high population density (usually urban areas) and in project areas. In specific situations, environmental manipulation or modification and house screening are other preferred methods. These methods are usually preferred in low-endemic areas, either alone or in combination with other methods.

(3) Biological control using larvivorous fish, biolarvicides, and methods that protect beneficial insects, e.g. in rice IPM: application of fish or biolarvicides requires comprehensive knowledge of vector ecology and geographical reconnaissance of breeding habitats. In confined water bodies, especially in arid and semi-arid ecosystems, use of fish has been found effective alone or in combination with other methods in an integrated approach.

Since detailed description on the criteria for selection of each of these components is beyond the scope of this document, further information on such criteria and procedures for selection of insecticides and other methods can be found in an earlier WHO publication (Najera and Zaim, 2002).[10]

Insecticide resistance monitoring and management

Public health use of insecticides has led to development of resistance in a number of vectors. Agricultural use of insecticides may also be involved in the buildup of resistance. For example, pyrethroid insecticides are widely used in agriculture; hence mosquito populations breeding in these agricultural environments on in water bodies affected through pesticide runoff from agriculture are likely to develop resistance to these compounds under the intense selection pressure occurring in such cropping systems. In this scenario, insecticide resistance management approaches may have to evolve through operational research to prolong the life of the limited arsenal of chemical insecticides.

Management of public health pesticides

It may be necessary to develop guidelines in local languages to educate public and relevant parties and build capacity for the sound management of public health and agricultural pesticides. Detailed guidelines are available on the web for the situation analysis[11] and for the sound management of pesticides.[12]

Intersectoral collaboration and partnerships

Intersectoral and often multisectoral actions are required for the application of preventive and control measures. This must be done in a coordinated manner to optimize the use of resources. Although national policies usually prescribe intersectoral coordination, institutional arrangements at district and sub-district level are often weak and non-functional for effective action. The following steps are therefore suggested for the VBD control programme:

---

[10] This document is available at WHO website using the following link:
http://whqlibdoc.who.int/hq/2003/WHO_CDS_WHOPES_2002.5_Rev.1.pdf
[11] http://whqlibdoc.who.int/hq/2005/WHO_CDS_WHOPES_GCDPP_2005.12.pdf
[12] http://whqlibdoc.who.int/hq/2003/WHO_CDS_WHOPES_2003.7.pdf and
http://www.fao.org/AG/AGP/AGPP/Pesticid/Default.htm)

- Identify stakeholders and potential partners within and outside the health sector.
- Identify the anticipated roles of each sector. Some of the possible actions are suggested in Table 3, but these are not exhaustive and may have to be built upon according to local needs.
- Establish a formal institutional coordination mechanism for monitoring the implementation and follow-up of actions.
- Organize regular formal and informal consultations with all key stakeholders to discuss relevant issues, provide feedback, strategic orientation, technical support and resources.
- Build up necessary technical capacity of the partner sectors wherever necessary to ensure subsidiarity and sustainability of desired actions within that sector.

Table 3: Anticipated roles of various sectors in IVM implementation*

S No.	Sector/agency	Roles
1	Agriculture	FFS approach to implement IPVM, popularizing the concept of dry-wet irrigation through extension education, pesticide management
2	Water resources development	Maintenance of canal system, intermittent irrigation, design modifications and lining of canals, weeding for proper flow, creating small check-dams away from human settlements, health impact assessment (HIA)
3	Water supply	Repair of leakages to prevent pooling, restoration of taps, diversion of wastewater to pond/pit, staggering of water supply, mosquito-proofing of water harvesting devices, repair of sluice valves.
4	Road and building sector	Proper planning as per by-laws, merging pits by breaking bunds, excavations in line with natural slope/gradient, making way for water to flow into natural depression/pond/river, follow-up actions after excavations.
5	Urban development	Implementation of building by-laws, improved designing to avoid undue water lodging, building use permission after clearance of health dept.; safe rainwater harvesting; use mosquito-proof design of dwellings; housing location.
6	Industry/mining	Improving drainage/sewerage system, safe disposal of solid waste/used containers, mosquito-proofing of dwellings, safe water storage/disposal, use of ITN/LLIN.

S No.	Sector/agency	Roles
7	Railways	Proper excavations, maintenance of yards and dumps and anti-larval activities within their jurisdiction; HIA for health safeguards.
8	Environment/ Forest	Pesticide management policies, environment management policies, reclamation of swampy areas, social forestry.
9	Fisheries	Institutional help/training in mass producing larvivorous fishes, promotion of composite fish farming schemes at community level.
10	Remote sensing	Technical/training help in mapping environmental changes and disease risk using GIS.
11	Private pest control agencies	Judicious use of insecticides, promotion of IVM-based sustainable preventive and control methods.
12	Planning departments	Involvement of health agencies at planning stage for HIA and to incorporate appropriate mitigating actions in development projects.
13	Sea/air ports	Vector surveillance and control measures.
14	Education	Developing training materials in local languages, school health activities incorporating vector control.
15	Mass media	IEC activities, advocacy.
16	Village Councils	Overall cooperation in the ongoing health programme and to ensure public participation as and when needed.
17	Local Govt.	Update public health by-laws.
18	Community	Household sanitation, use of ITN/ LLIN, acceptance of IRS.
19	NGOs	Community mobilization, village-level training, distribution of IEC material, ITN promotion, orientation of architects/builders.
20	R&D industry	Development of new, safer and more effective insecticides/formulations; promoting safe use of public health pesticides.
21	Health (Vector-Borne Disease Control Programme VBDCP)	Lead sector to develop IVM guidelines, conduct situation analysis and VMNA; plan, implement, coordinate, guide, monitor and evaluate IVM activities; operational research; capacity building activities; advocacy; resource generation.
	Health (non-VBDCP)	Promoting ITN/LLINs through health and family welfare services, integrated management of childhood illness (IMCI)

*The list may be expanded according to local needs

Community participation

Empowerment of grassroots communities is necessary to ensure acceptability/ sustainability and better effectiveness/cost-effectiveness of interventions. It is important to consider several issues regarding empowering communities. These will include a participatory process to:

- identify the local communities which will play a key role;
- conduct special social surveys to assess community knowledge, attitudes, beliefs and practices (KABP);
- determine the roles of local communities and individuals;
- decide on approaches/methods to develop community participation;
- assess mutually beneficial actions to determine sustainability of actions;
- assess resources and technical support required

Advocacy and social mobilization

There are various ways to involve communities. One approach successfully developed by the agricultural sector is IPM/IPVM.

Farmer Field School (FFS) approach, wherein farmers become experts in pest and vector control through a participatory learning approach to deal with local situations (see Box 1). The other approaches are through information, education and communication (IEC) campaigns, behaviour change communication (BCC) for vector control and participation by nongovernmental organizations (NGOs)/community organizations.

Important steps include developing advocacy, sensitization and participatory materials for different levels of implementation. At community level, identification of prime movers and community leaders is also necessary.

Advocacy efforts will be required to create an enabling policy environment for policy- and decision-makers as well as collaborating sectors to allocate more resources for preventive actions against vectors and accept an IVM approach. This is all the more important since policy decisions in favour of IVM, and mobilization of human and financial resources by public sectors, will involve different departments at district level and different ministries at state/provincial and national levels.

Capacity building

The health sector capacity to conduct routine entomological surveillance and insecticide resistance monitoring, as well as plan, implement and monitor/evaluate vector control, is inadequate. Capacities are weaker at district level. Implementation of IVM therefore requires building programme capacities at district level, though some of the activities require greater entomological skills which can be built/strengthened at higher levels (such as zonal/state/municipality/provincial). Some actions will be required at the national level, such as integrating IVM principles and decision-making criteria into the existing framework of national health policies.

Capacity-building will include actions such as:

- building health/VBD control sector capacity through provision of an enabling environment, e.g. political commitment, development and implementation of appropriate recruitment and career policies, provision of facilities and resources, strengthened training institutions;

- establishment or strengthening of vector control services in conjunction with:
  - creation of enabling policy framework for intersectoral actions;
  - collaboration to strengthen legislative and relevant regulatory frameworks and their enforcement;

- building institutional capacity:
  - through intensification of training and re-training of personnel to develop technical, financial and programme management and cost-effectiveness measurement skills; skills for curriculum development; piloting and standardization of training materials and evaluation of training process;
  - through improved provision of infrastructure;

- building capacity of other sectors outside health sector as well as of the private sector such as pest control operators, architects'/builders' associations, NGOs, etc.;

> strengthening of available capacity at district level to manage public health pesticides as per standard guidelines;

> technical support mechanisms, e.g. information, communication and supply systems to support trained personnel, supervision, monitoring and evaluation;

> where indicated and feasible, support for strategies for community empowerment, as exemplified by the engagement of farmers in integrated pest and vector management;[13]

> training of operational staff on IVM as a continuous process.

Monitoring, Evaluation and Operational Research

Monitoring and evaluation are essential components of IVM. Monitoring measures the implementation of its range of activities (the process), while evaluation measures the extent to which direct outcomes have been achieved. Impact assessment determines the effects or the impact attributable to the programme. The inputs and processes required to deliver each activity or intervention, and their relative contribution to the overall impact, must be assessed for effectiveness, cost-effectiveness and sustainability in a given situation. In Table 4 below, a scheme is suggested to identify expected results and decide indicators for monitoring and evaluation, which in turn will require development or use of available methods/tools.

Table 4: Log-frame for planning vector control using an IVM approach.

Expected results	Activity	Objectively verifiable indicators	Method of evaluation	Assumptions/ risks	Resources required
1.		1. Process indicators			
		2. Outcome indicators			
		3. Impact indicators			
2.					

---

[13] see more details at: http://www.searo.who.int/EN/Section23/Section1001/Section1110_12796.htm

Involvement of partners and community representatives in participatory evaluation is also important, because it increases programme ownership and has potential to generate data on behavioural, social and political changes that would be difficult to obtain thorough interviews.

Operational/implementation research

A number of issues will need scientific examination to develop feasible, cost-effective, socially acceptable and thus sustainable interventions for each local eco-epidemiological setting/stratum. Monitoring and evaluation can be considered a part of operational research in the context of IVM since the outcomes will enable improvement of inputs and implementation processes. The operational research issues will be identified for each district. Some of the key areas may include KABP surveys to determine community acceptance of interventions, evaluation of effectiveness of IVM programme, insecticide resistance monitoring, evaluation of new vector control intervention methods, etc.

3.8 Budgeting

Like any other plan, the district IVM implementation plan also will require estimating the resources required and then preparing a budget estimate covering all possible anticipated activities and keeping in view the timeframe.

4. References

(1) Berg, Henk van den et al. (2007). Reducing vector-borne disease by empowering farmers in integrated vector management. WHO Bulletin, 85 (7): 501-568.

(2) Food and Agriculture Organization of the United Nations (FAO) (2002). International Code of Conduct on the Distribution and Use of Pesticides, Revised Version adopted by the Hundred and Twenty-third Session of the FAO Council in November 2002

(3) Najera J.A. and Zaim M. (2002). Malaria Vector Control. Decision Making Criteria and Procedures for Judicious Use of Insecticides. WHO/CDS/WHOPES/2002.5, WHO, Geneva.

(4) WHO (1983). Integrated Vector Control. Seventh Report of the WHO Expert Committee on Vector biology and Control. WHO Technical Report Series 688. World Health Organization, Geneva.

(5) WHO (1993). A Global Strategy for Malaria Control. World Health Organization, Geneva.

(6) WHO (1995). Vector Control for Malaria and Other Mosquito-borne Diseases. WHO Technical Report Series 857. World Health Organization, Geneva.

(7) WHO (2004). Global Strategic Framework for Integrated Vector Management. World Health Organization, Geneva.

(8) WHO (2005). Regional Framework for an Integrated Vector Management Strategy for the South-East Asia Region. Document SEA-VBC-86. World Health Organization, Regional Office for South-East Asia, New Delhi.

(9) WHO/SEARO (2006). The Revised Malaria Control Strategy, South East Asia Region: 2006-2010. Document SEA-MAL 243. World Health Organization, Regional Office for South-East Asia, New Delhi.

(10) WHO/SEARO (2006). Report of the Regional Workshop to Implement Integrated Vector Management (IVM), Vector Control Research Centre – VCRC, India, 18th to 21st December 2006, Puducherry and Tricchy, Tamil Nadu, India.

5. List of Contributors

(1) Prof. A.P. Dash, Director, National Institute of Malaria Research, Delhi

(2) Dr Rajpal Yadav, Deputy Director (Senior Grade), National Institute of Malaria Research, Field Station, Nadiad, Gujarat

(3) Dr Henk Wendenberg, Consultant, F.D. Rooseveltstraat 29, 9728RV Groningen, The Netherlands

(4) Dr Sabesan, Deputy Director (Senior Grade), Vector Control Research Centre, Puducherry

(5) Dr Jambulingam, Deputy Director (Senior Grade), Vector Control Research Centre, Puducherry

(6) Dr N. Arunachalam, Deputy Director (Senior Grade), Centre for Research in Medical Entomology, Madurai

(7) Dr S.K. Ghosh, Deputy Director (Senior Grade), National Institute of Malaria Research, Field Station, Bangalore

WHO:

(1) Dr Kazuyo Ichimori, Scientist, Vector Ecology and Management, Department of Control of Neglected Tropical Diseases, WHO/HQ

(2) Dr Chusak Prasittisuk, Coordinator, Communicable Diseases Control, WHO Regional Office for South-East Asia

(3) Dr Krongthong Thimasarn, Regional Adviser-Malaria, WHO Regional Office for South-East Asia

(4) Mr Alexander von Hildebrand, Regional Adviser-Food and Chemical Safety, WHO Regional Office for South-East Asia

(5) Dr Abraham Mnzava, Regional Adviser, Vector Borne Disease Control, WHO Regional Office for the Eastern Mediterranean